ESPIONAGE BLACK BOOK THREE

In this series:

Espionage Black Book One: Intelligence Databases Explained

Espionage Black Book Two: Codes and Ciphers Explained

ESPIONAGE BLACK BOOK THREE:
Surveillance Explained

Henry W. Prunckun

Bibliologica Press

Espionage Black Book Three:
Surveillance Explained

ISBN 978-0-6452362-0-0

A catalogue record for this
book is available from the
National Library of Australia

For information on all Bibliologica Press's publications,
visit our Web site at bibliologica.com

Bibliologica Press
P.O. Box 656
Unley, South Australia, 5061
Australia

CONTENTS

— CHAPTER ONE —

SURVEILLANCE

Isobel was sitting in the backseat of her car, just behind the driver's seat. Nonetheless, she had a clear view of the townhouse diagonally across the street.

It was early—5:45 am. Although it was a warmish autumn day, it was chilly. Yet, the thought of leaving the engine running to heat the interior didn't cross her mind. To do so was tantamount to placing a sign on her vehicle saying, "Look, I'm here." That's why she sat in the rear, so anyone looking at the car from the front would see an empty vehicle.

Isobel held her D-SLR camera on her lap and passed the time by listing to the local AM station that read the newspapers for the vision impaired.

At 6:00 am, the front door opened, and a man fitting the description of her target appeared. He turned, inserted a key in the door's deadlock, and then made his way to a stellar-blue Mustang that was parked in the reserved space in front of the townhouse.

"Izzy" raised her camera and could see his face clearly through the camera's telephoto lens as he entered the car, so she snapped a few images. She was careful to include the car's registration and the row of townhouses in the photo's background. *The mission is on*, she thought.[1]

---

1. In this book, the term *mission* is synonymous with an *operation*.

She followed him through the morning traffic, varying her place in the line of cars behind him and changes lanes to avoid becoming a feature in his rear-view mirror. This "close tail" was filmed with her dash camera.

She broke contact when he entered a gym about fifteen minutes later. She surmised that the gym was his part of his daily routine. So, the next day, Isobel set up her observation post in the gym's parking lot and waited for him to appear, which he did at 6:15 am.

Figure 1—Using a vehicle as a fixed observation post. Courtesy of the author.

She exited her car at the same time he did and followed him into the health club. All along, Izzy filmed him with a concealed camera that was hidden in the shoulder strap of her backpack. Later, her video showed the target walking without the limp he had said he incurred in a worksite mishap. Her footage showed him running on the treadmill as she passed the large display windows that faced the public area.

The next day Isobel submitted her surveillance report via email to the insurance company that hired her. It was accompanied by the digital images and videos she captured. She never heard any more from the claims manager about the mission because it was likely that the claimant in the Mustang withdrew his petition for an insurance payment.

This type of scenario is not limited to private investigators. Police detectives, government investigators, espionage agents, intelligence officers, and covert military operatives all use surveillance.

Surveillance is observation.[2] But it is more than watching. It refers to observations that are discreet with the purpose of obtaining information. The fact that these observations are discreet implies that it is a covert activity, and this is true.

If we look at surveillance's typology, we see that it comprises several classifications. Each class is determined by the context in which it is used. In this book, we will discuss the classifications of static (i.e., fixed or the stake-out), foot, mobile, rural, aerial, and technical surveillance.

Our examination of surveillance will also take in a study of some key pieces of equipment used in surveillance, as well as looking at anti-surveillance and countersurveillance measures. Before we do that, let us first look at how operatives plan and prepare for missions.

Every mission begins with an objective. This is sometimes called the *terms of reference*. This can be a simple document or a detailed set of instructions. For the mission that our notional private investigator Isobel completed, her objective might have been to "Obtain

---

2. Henry Prunckun, *How to Undertake Surveillance and Reconnaissance* (South Yorkshire: Pen & Sword Books, 2015).

photographic evidence that either supports or refutes the assertion that he (the target) is fit to return to work."[3]

The terms of Izzy's reference might seem obvious, but human nature being what it is, an investigator could be tempted to snap photographs of everyone entering and exiting the target's house looking for possible other illegal activity. This would be a waste of time, and lost time means lost income.

All surveillance activities need to be funded; unlike any impression gained from television shows or movies, conducting a surveillance mission is expensive. Someone must authorize the mission, and that person is responsible for paying for it. Hence the need for the scope to be set before initiating the mission—what is "in-scope" must be defined. Everything else is "out-of-scope" unless something peculiar develops during the mission, but even then, permission needs to be received if the mission's objective is to be widened to include new activity.

In sum, a well-stated surveillance mission objective sets the scope, which outlines the parameters of the investigation. Therefore, like a compass, it helps the mission stay focused and alerts the operative(s) to when they have achieved their goal.

---

3. Note the words, "...supports or *refutes*..." The lack of information, or information that points to a contrary position, is as important as evidence that supports the hypothesis. One need look no further that the biased information put forward in the 2003 by the then George W. Bush's Administration's fallacious case of Iraq's weapons of mass destruction. See, for example, Scott Ritter, *Iraq Confidential* (New York: I.B. Tauris & Co, 2005), and Valerie Plame Wilson, *Fair Game: My Life as a Spy, My Betrayal by the White House* (New York: Simon & Schuster, 2007).

Plans can be either long term or short term. The former is considered strategic and the latter operational or tactical.[4] It does not mean that once a surveillance plan is created, that is the end. Like all plans, unforeseen events occur when unexpected. So, plans can change. If, for instance, a new activity is observed during a stake-out, a new tactical target can be included in the strategic plan *if* authorization is received.

One way surveillance missions are planned is by using a method employed by the military, law enforcers, and emergency services—SMEAC (pronounced *smee ack*). SMEAC is a mnemonic for situation, mission, execution, administration and logistics, command, and communications (or signals).[5] It is easy-to-remember format was designed to layout the important aspects of a mission:

- Situation (A summary of the current situation);
- Mission (A single sentence statement about the task's objective or what needs to be done);
- Execution (The plan to accomplish the mission. Although presented in simple terms, it needs to contain enough details so that operatives carrying out the surveillance can do so successfully. This aspect of the

---

4. "Tactical intelligence provides support for an operation that is either under way or about to begin." … "Operational intelligence is information that contributes directly to the achievement of a wider goal, whereas strategic intelligence relates to long-term forecasts or broader conclusions on larger objectives." Hank Prunckun, *Method of Inquiry for Intelligence Assessment, Third Edition* (Lanham, MD: Rowman & Littlefield, 2019), pp. 15 and 16)

5. The following has been adapted from my book, Prunckun, *How to Undertake Surveillance and Reconnaissance*, pp. 96–98.

SMEAC may contain visual aids such as diagrams, maps, and overlays);

- Administration and Logistics (The composition of the surveillance team that will be involved, how the plan will be executed, what the arrangements will be for food, water, fuel, sleeping, ablutions, equipment, as well as the paperwork for logs and other recordkeeping requirements); and
- Command and Signals (A clear summary of who is in command and the legal authority for carrying out the mission. Directions need to be given as to how and when the surveillant(s) will communicate, e.g., the two-way radio channels or frequencies, cell/mobile telephone numbers, code words to be used for predefined situations, etc.).

Having received a SMEAC briefing, which can be audio–visual presentation or a one-page written brief, the surveillant(s) need to assure themselves that they understand:

- The mission's objective;
- The task(s) that needs to be accomplished;
- If more than one surveillant, each person's role during the mission; and
- The timeframes and criticality of the mission.

The literature of training shows that the best way these four points can be accomplished by what is called "active listening" (i.e., re-stating or paraphrasing what you have heard for clarity), [6] taking notes, and asking questions if

---

6. Harry Weger Jr, Gina Castle Bell, Elizabeth M. Minei, and Melissa C. Robinson, "The Relative Effectiveness of Active Listening in Initial Interactions," *International Journal of Listening*, 2014, Volume 28, Number 1, pp. 13–31.

any aspect of the mission is unclear, or raises concerns about safety, legality, or ethics.

Before we move on to our discussion of the different classifications of surveillance, it is worth mentioning the difference between surveillance for evidentiary purposes and surveillance for gathering intelligence. *Evidence* is proof of something. It can be physical evidence, as in the type of data crime scene examiners collect, or it can be eye-witness accounts of events or conversations. Usually, this type of information manifests itself in some verifiable proof—laboratory assays, photographs, biological material, and so on.

In contrast, *intelligence* is information that may be evidentiary in nature but can be information that is in some way related to the issue under investigation. It can be tangential information or information that is indirectly or consequential associated with some happening. So, intelligence information does not carry the same standard of proof as evidence. However, its purpose is not evidentiary; these data are combined with other data by analysts to reduce uncertainty.[7]

Intelligence provides a way to interpret the issue under study so that resources can be tasked to, perhaps, then collect evidence. Intelligence analysis is an academic discipline that focuses on methods that help reduce ambiguity in uncertain situations.

---

7. Prunckun, *Method of Inquiry for Intelligence Analysis, Third Edition*, 2019.

— CHAPTER TWO —

STATIC SURVEILLANCE

Throughout the Cold War, surveillance was a prominent feature in many fictional accounts of espionage and clandestine operations. Since 9/11, the threat of Communism as a trope has been replaced by terrorist groups, computer hackers, and organized criminal networks that traffic drugs, arms, and people. These issues reflect the current global worries. Nevertheless, we see surveillance still being portrayed in stories about counterintelligence, espionage, and special and psychological operations targeting these threat agents.

It seems that many of these accounts lever the technological advances[8] that have transformed the way surveillance was once carried out as the drawcard for "selling" the story. Yet, as will be explained, starting with this chapter, although new devices are now aiding surveillants, surveillance's first principles[9] have not changed.

Physical surveillance, or more precisely, static physical surveillance, is the act of observing people, vehicles, or activities from a fixed position. This is sometimes referred as a *stake-out*.

A stake-out is an effective means of obtaining information about the target's movements and habits as

---

8. Especially regarding electronic and ariel drone technology.

9. Although somewhat pejorative, the label *gumshoe* came about for a reason. This is because surveillance relies on investigators pounding "shoe leather" on pavements.

well as the people they associate with, which can infer their reputation and character (i.e., intelligence). It is also an effective means of collecting evidence (i.e., proof).

Although espionage fiction tends to glamorize many of the aspects of spying, one facet it does not is that of fixed surveillance. That is because it is mundane, tedious, and boring, except when the target is being viewed. Then, the surveillant must be alert and able to respond to situations that may not have been planned.[10]

It is a paradoxical activity. On the one hand, it is a sedimentary activity where hours or days could be spent sitting and watching with nothing happening. On the other, at perhaps the most unpredictable time, the target could appear, so the surveillant must shift to full attention, ready to act.

Setting up an observation post can take imagination because it is essential that the surveillant does not stand out. They must blend in the surroundings like a chameleon does in the wilderness. A parked car might be the most typical "blind," but the situation might call for using a delivery truck or other vehicle. It might mean viewing the target or area from a greater distance so that anyone looking for a surveillant would not notice them through a narrow opening between, say, two buildings, a distant window in a public building, or as a musician busking outside a café.

In the case of the latter, fixed surveillance calls for imagination to develop disguises. If the surveillant is to presents as a musician, then they need to look the part with a guitar case, portable amp, and if necessary, the ability to play. This is demonstrated in the following case study.

---

10. Chuck Chambers, *The Private Investigator's Handbook* (New York: Perigee Books, 2005), pp. 193–205.

After waiting several hours in his parked car opposite a public park, a police surveillant's target and another person of interest arrived for what was anticipated to be a meeting to discuss their illegal enterprise. But, the surveillant realized that he needed to move from his location because it would immediately become obvious to the targets that his presence was odd.

Figure 2—Fixed surveillance calls for imagination to develop disguises that go beyond the typical CCTV security camera installation (pictured). Courtesy of the author.

So, he exited his vehicle and made somewhat of a show of opening his car's truck and retrieving some pamphlets and a small portable stand. He set up the table and banner sign at the park's entrance where he could have an unobstructed view of the two sitting at a nearby picnic table.

Once the targets read the sign, which stated he was handing out religious material, they took no further notice. Yet, from his observation post, the surveillant was able to capture the whole encounter via a digital camera hidden in a book sitting on the stall's collapsible table.

⌘ ⌘ ⌘

In the case where a building is under observation because there may be people of interest coming and going, a more permanent observation post is necessary. Such a place might be a dwelling, office, or storage accommodation that has unobstructed line-of-sight to the target area—say, the target's door, parking lot, or workshop, etc.[11]

In such a case, two or more surveillants can be tasked to rotate observation duties, thereby relieving eye strain and mental fatigue from constant concentration. This also allows the observer to call out the activities that are being noted, with their partner recording the details and communicating instructions to a command post when necessary. This arrangement also permits for around-the-clock shifts.

Tradecraft[12] plays a large part in how the observation post operates. The sight of a long-lens camera protruding from a window indicates an amateur operation. Curtains or other window treatments need to be considered for daylight observation. These treatments need to facilitate obscuring detection by passers-by yet allow unincumbered viewing of the target. Night observations must consider the elimination of backlighting that may silhouette operators or show movement.

The typical surveillance van that is portrayed in cinema and television is used, but more sophisticated observation posts can be made from any type of vehicle by making them remotely operated. For example, a car can be

---

11 In one case, private investigators rented a large trailer-towed sailing yacht and parked diagonally across the street from the target's house. Two PIs hid inside and recorded evidence that supported the allegation the target lodged a false disability claim with his insurer. The boat's cabin had a small galley and toilet facilities as well as a bunk for rest during rotations.

12. i.e., All practical skills relating to working in the fields of intelligence, espionage, counterintelligence, and covert operations.

parked, and using, say, a modified dash-cam (e.g., pan-tilt-zoom) and a transmitter in the car's trunk, surveillants can make their observations of the target and surrounding area. This has many advantages, especially eliminating the stress of having to sit for hours in a vehicle without a latrine.

Figure 3—Traffic camera mounted on a signal pole at a city intersection.

Used alone, or in combination with traffic cameras (which are usually mounted over intersections), a mobile surveillance operation can be conducted in a way that the target will not be able to detect, because no one is physically following them—this is, in effect, a series of fixed observation posts.

— CHAPTER THREE —

FOOT SURVEILLANCE

Foot surveillance is arguably the most demanding form of surveillance. This is because it relies on the surveillant's instincts rather than on technical aids. Imagine for a moment the scenario where a surveillant is tasked to follow a target for hours. It becomes immediately clear that it will take great skill to do this without being detected, especially if the target is security conscience. Therefore, foot surveillance is seen as an art, not a science, though, like surveillance in general, planning and tactics play their part.

Foot surveillance is referred to as *tailing,* and there are three types of tails—the close tail, the loose tail, and the rough tail. The choice of each type of surveillance is determined by the mission. A close tail is used where the target needs to be observed, their actions monitored, and not lost. It is a high-risk operation, but if the target is, say, suspected of servicing a dead drop,[13] then anything the target touches or any person they meet may be critical in understanding the issue under investigation.

A loose tail is used where it is more important not to be seen. In situations where it is crucial that the target does not discover they are being watched, then observing from

---

13. A *dead drop* can be either a place or a receptacle that is used to by operatives to exchange messages, documents, or drives containing digital files. A dead drop obviates the need for operatives to meet in person.

a distance is a clear choice. Though, it also means that there is a risk of losing the target.

The rough tail is one in which detection is not a central concern. In this type of tail, the target may become aware that they are being followed, but the mission does not require secrecy. In fact, the rough tail may be used to communicate to the target that they have been discovered, and whatever activity they are involved in is known to the investigating agency.[14]

Planning is valuable but may not always be possible. Nevertheless, surveillants will follow a structured approach to organizing their mission. The consideration is to understand the geographic area in which the target will be traversing. It is self-evident that different locations require different physical appearances—a smart up-town urban area requires dress and manners that will be different to those required in a regional blue-collar town or a suburban setting.

Understanding of the geography is guided by what is known of the target—background intelligence. Knowing their what they do for work, personal interests, habits, likes-and-dislikes, friends, family, and places they enjoy going all assist in planning.

Coupled with the first two considerations is having prepared for physical appearance—change of clothing and disguises. This can be as simple as removing or adding an article of clothing, putting on or taking off a hat, adding or removing sunglasses, and one of the most overlooked items, a change of shoes. A surveillant can change their overall appearance relatively efficiently, but experience

14. An agency may even want to disrupt the target's operation by overtly preventing the target from meeting with their agent.

operatives have pointed out that shoes are the weak link in disguise the chain.[15]

Money and credit/debit cards are the next item for inclusion in any foot surveillance plan. Surveillants need to be prepared to jump on a bus, train, subway, or taxi should the target do the same.

The fact that the target might move from one locale to another during the follow means that the surveillance team needs to have anticipated such a contingency. Moreover, it needs to be planned what will happen if there are signs of a "handoff" or a pick-up of some kind. Questions like, will this type of encounter requires one of the surveillance team to break and pursue this potential target? Who will that be? How will this be communicated to the group?

On the question of communication, operatives will need to be briefed on whether hand signals or radio/telephone communications will be used. If the latter, the type and concealment, or disguise, will need to be thought through, as will whether there will be provision for backup surveillants if an operative needs a rest or to rotate a surveillant to avoid detection.

Figure 4 shows a typical surveillance deployment involving a team of three operatives. It comprises two surveillants following the target from the rear and one following in parallel on the opposite side of the street. For clarity, the diagram does not show other pedestrians, nor the vehicles parked along or traversing the road.

---

15. ACM IV Security Services, *Secrets of Surveillance* (Boulder, CO: Paladin Press, 1993), p. 18. This especially true if the surveillance team is using the *waterfall* technique where operatives walk directly toward the target rather than follow from behind. Robert Baer, *See No Evil* (New York: Crown Publishers, 2002), p. 42.

The reasoning behind this formation is that it offers the team the versatility to change positions as the tail progresses. Take, for example, Figure 5, where the target tuns right at the corner of the street. If the target is using anti-surveillance methods, they may be alert to the same people around them. However, the diagram demonstrates how surveillants can change position.

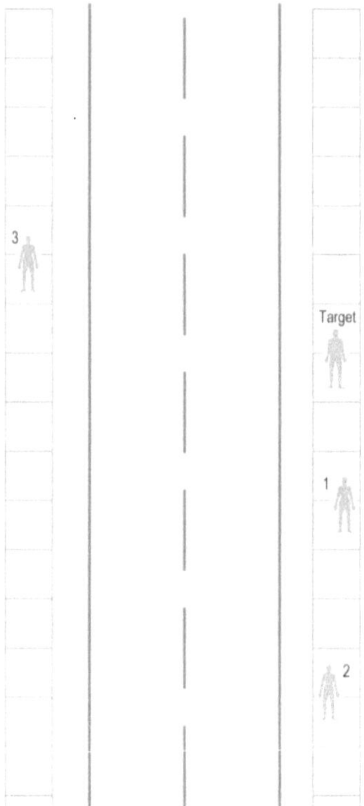

Figure 4—A typical foot surveillance formation.

Figure 5 shows operative 1 swapping roles with operative 3 by crossing the street. Operative 2 comes forward to replace operative 1, and operative 3 takes over operative 2's position. This type of manoeuvre can be made seamlessly, and as such, would be difficult to detect by the target.

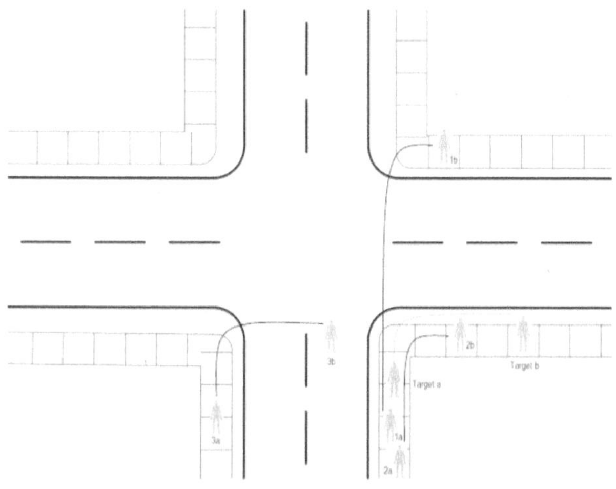

Figure 5—An example of how surveillants switch positions.

Other manoeuvres that cater for varying actions of the target are possible. Variations of these moves are also possible for teams of less or more operatives, including stationary (static) surveillants placed at a strategic position along routes the target might take (see Chapter Four regarding "progressive surveillance"). The latter allows for the team to "disappear" for a period (perhaps to change disguises) and reappear once the target is on the move again.

Having a reliable radio communication network that links operatives with each other as well as their command center (if one is being used), the team can react quickly to changes the target might take that were beyond anticipating.[16] Another aid for quick reaction is the ability to conduct a tail using a vehicle, which we will discuss in the next chapter.

---

16. Michael Chesbro, "Clandestine Communications Methods," in Hank Prunckun, *Intelligence and Private Investigation* (Springfield, IL: Charles C. Thomas Publishers, 2013), pp. 145–158.

— CHAPTER FOUR —

MOBILE SURVEILLANCE

Vehicle surveillance allows for swift maneuvering, but by its nature—people on the move—it is a high-risk tailing method. This is because the target can confound the surveillance detail—either intentionally or inadvertently—through traffic conditions (e.g., heavy traffic), road peculiarities (e.g., merging lanes or road repairs/construction/detour), or bad luck (e.g., sudden changes of traffic signals, or having to stop as an emergency vehicle passes).

Analogous to the disguises a foot surveillant uses, the mobile surveillant needs their vehicle to blend in with the locale. This is an important facet of shadowing because the type of vehicle will not be the same for each operational tasking. For instance, in a rural setting, which vehicle would most likely fit in: 1) a Mercedes-Benz; 2) a small economy car; or 3) a pick-up truck?

Likewise, of these three vehicles, which would be less likely to be noticed in a middle-class suburban neighbourhood? And, what car might be customarily seen driving around a neighborhood where the average income is six-digits?

Vehicles, and their on-road costs and maintenance, are expensive. Therefore, rentals immediately present as a viable option for small surveillance units.[17] Cars, trucks,

---

17. If not an outright alternative, this arrangement is certainly a consideration to supplement an agency's pool of surveillance vehicles.

and vans can be changed daily or even during the day if required. This versatility is not likely if an agency maintains its vehicle fleet.

⌘ ⌘ ⌘

Some television shows depict surveillants parked in cars down the street from the target's premises. In these shows, the target is seen leaving, say, their home, not noticing the surveillance vehicle as they drive off; nor do they see the surveillance vehicle immediately start-up and follow them. Is this portrayal realistic? No, of course not.[18] As former Chief of CIA Counterintelligence stated, "Surveillance is excruciatingly difficult. ... [TV style] surveillance would be detected in a heartbeat."[19]

How might this be done so that the chance of the surveillant being "blown" is reduced? Well, if some sort of "trigger" could be deployed that alerts the team to when and in which direction the target is heading would be effective in fooling the target.

If we brainstorm a few possibilities, we can conceive a situation where an observation post is set up some distance away, allowing "eyes on the target." Observers there could radio the details to the mobile team who would be parked in the vicinity. A variation of this might involve an action camera or wildlife/trail camera with a wireless connection to the Internet. This device could act in place of a person sitting in a car.

Once the tail has been initiated, experience becomes the best guide in determining how to conduct the follow.

---

18. Unless the vehicle is disguised, as in the case study cited where the surveillants used a sailing yacht. See footnote 11.

19. James M. Olson, *To Catch a Spy: The Art of Counterintelligence* (Washington, DC: Georgetown University Press, 2019), p. 45.

Nonetheless, there are theoretical assumptions that suggest certain actions, but it is the surveillant's experience that governs whether the tail will be successful or not.

These theoretical aspects are taught in surveillance courses, such as where to position the surveillant's vehicle in relation to the target. Figure 6 suggests three such positions, each offering cover by using other vehicles as a screen.[20]

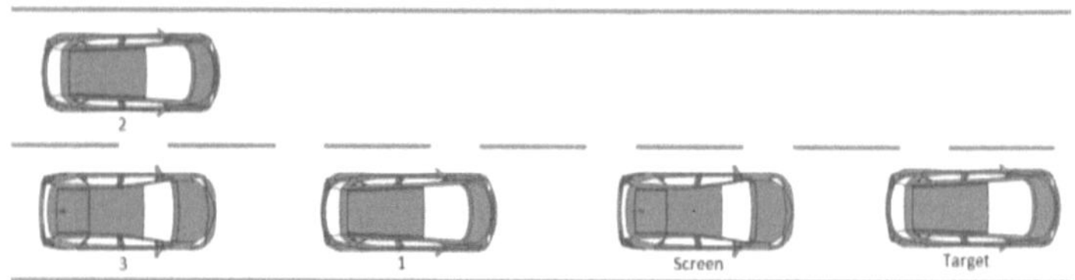

Figure 6—Three places that allow for cover in traffic.

Will such advice work in all situations, locales, and with all targets? No. But these words of advice form what is a suite of standards termed *first principles*. That is, the foundational thinking behind good operating practice. Although we cannot discuss the encyclopedic knowledge gain by operatives through years of experience, we can review some of the tenets of good operating practice. The first, and most obvious, considers adapting to traffic conditions.

As road and traffic conditions change, so will the tactics employed by the mobile surveillant. For example, if traffic thins out providing less screening, the surveillant can change their posture, and hence their profile, within

---

20. If the surveillant uses either position 2 or 3, then position 1 would also be a screening vehicle.

the car.[21] Putting on a hat, taking off a hat, or if there are with a partner, the partner can lay down in the back seat.

Again, experience, as well as imagination, play key parts in how well a target can be deceived. This tenet suggests that the surveillant visualize how the target will view those around them and then adopt the least obtrusive profile possible.

Another practice is to tail a target by following them in parallel or progressively.[22] Figure 7 shows a parallel formation, whereas Figure 8 demonstrates how a progressive tail could be set up.

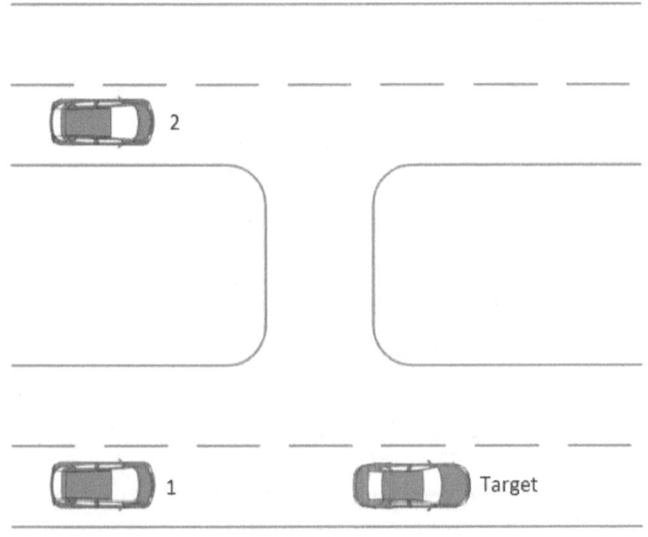

Figure 7—Example of a parallel follow.

---

21. For example, slumping in the seat or sitting higher or straighter.

22. There are, of course, other methods such as the *waterfall* technique discussed in footnote 15, but these comprise the most used tactics.

Although this Figure is set out in a series of uniform city blocks, we need to acknowledge there are a variety of variations. Regardless, the parallel concept shows how other vehicles in the surveillance team can engage with the target without being seen. As with the tail shown in Figure 7, the two pursuing vehicles can switch positions at, say, logical spots, like intersections.[23] In this regard, radio or telephone communication is vital. A GPS device allows surveillants to visually survey the road network to plan handover points during the follow.

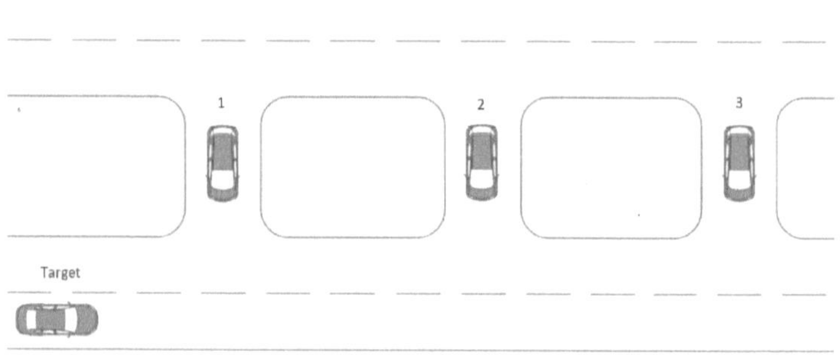

Figure 8—Example of progressive surveillance.

Progressive surveillance relies on several vehicles positioned at strategic points along the target's route (see Figure 8). As the target passes each waypoint, the surveillant at that point takes up the position of "command."[24] Of course, this type of formation assumes that the start and end points are known.

23. One drawback of the parallel follow is that there will be times where the target might not be under surveillance. This is likely to happen as surveillants break and another joins.

24. *Command* refers to the surveillant who has primary responsible for having "eyes-on-the-target." The brevity code word for command is *zero-zero*. ACM IV, *Secrets of Surveillance*, p. 225.

If the route is not known, then a variation of this method can be used. This technique calls for the surveillant to follow the target for a reasonable distance, then break contact. The next day, the surveillant positions themselves out of sight at the point where contact was lost the day before and waits for the target to pass. The target is then followed a reasonable distance, and again, contact is broken. This pattern of progressively following the target can be repeated until the target's destination is discovered.[25]

Although this method offers a great deal of protection from discovery, it is limited to situations where the target habitually follows the same itinerary. If the target is involved in illegal activity, is conducting confidential government work, or is a business person engaged in research and development, etc., then it is likely that they will be employing anti-surveillance measures, such as taking a different route each day, circling back, taking an indirect or convoluted route, or using a surveillance detection route (SDR) to expose surveillance (see Chapters 8 and 9 for a discussion of these techniques). Therefore, in many cases, this method will be inadequate.[26]

There are numerous variables that affect the success or failure of a surveillance operation. Apart from the target executing an ingenious anti-surveillance maneuver, there are other vehicles on the road to contend with, traffic lights, bends in roads, road construction, and car

---

25. A hypothetical example of this was discussed at the start of Chapter One.

26. Espionage is full of uncertainties. If a government operative were to demonstrate a habitual pattern of travel, as would be required for this technique to succeed, they might be aware of surveillance and allowing their pursuers to practice their progressive observation to lead them to a false location.

breakdowns, to mention a few hazards. With all these obstacles, it is very likely that there will be a break in contact (known as *lost command*) where there no one is able to see the target. When this happens, surveillants will put in place what is termed the *lost-command drill.*

The lost-command drill entails as series of actions that will help the team find the target. These steps do not ensure that they will find the target because there are too many confounding travel irregularities but will help reduce faulty search logic and guesswork.

The first action is for the command vehicle to advise the team that the target is not in sight ("temp unsighted"). The command vehicle will then try to reestablish contact by moving ahead, changing lanes, or whatever is necessary to get a clear view of the road ahead. If these actions fail to spot the target, the command vehicle will signal that contact has been lost. The command vehicle will take the most logical route, and the other cars in the follow will take the next most logical route, and so on until there are no more vehicles available. Figure 9 shows a simple example of this division of search tasks.

In this diagram, the command vehicle proceeds straight, leaving the other team members to go right and left. Each will continue until the target is sighted or they will turn back based on an estimate that uses the time-distance method of reckoning. This formula is expressed mathematically as: Time travelled x Speed of vehicle = Distance (T x S = D).[27]

By way of example, if the time since contact was lost was two minutes and the target's speed was forty miles per hour, then the distance where the target is likely to be sight

---

27. This formula can also be expressed as Distance x Speed = Time (D x S = T), and Distance x Time = Speed (D x T = S).

would be around two miles. Here is the result of the computation:

- T x S = D
- 2min/60min x 40mph = D
- 0.03 x 40 = 1.2 miles

In countries that use the metric system, this formula also applies; the only difference is that the speed and distance will be expressed in kilometers per hour and kilometers, respectively.

Does this mean that surveillants pull out a pad-and-pencil to do this estimate? No, this method is a guide. It suggests when it is safe to assume that the route they are travelling is not the one used by the target. Because the method only involves two factors (time and speed), it is easy to calculate in one's head while driving.

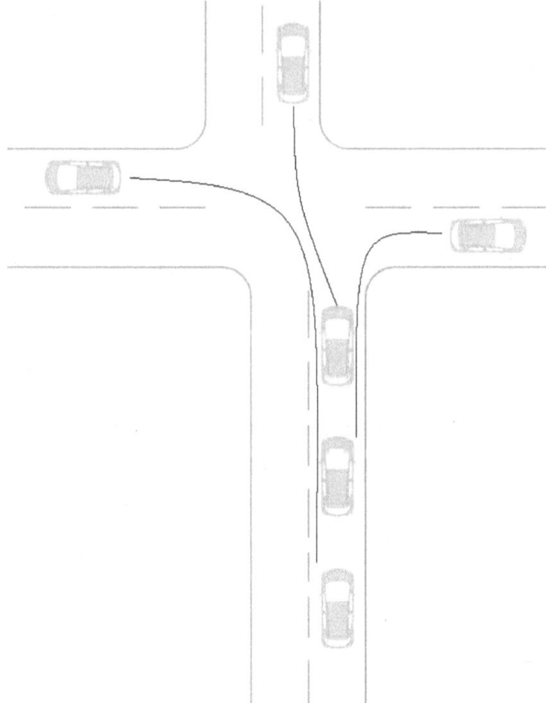

Figure 9—Lost-command drill.

— CHAPTER FIVE —

AERIAL SURVEILLANCE

Aerial surveillance by rotary and fixed-wing aircraft is an effective method of making observations. The history of aerial surveillance dates from the mid-19th century when the first photograph was taken from a French military hot air balloon.

Since that time, many developments have occurred in aerial surveillance. Technology that is currently available for airborne surveillance ranges from secret reconnaissance satellites and ultra-sophisticated spy planes, down to light aircraft and helicopters with conventional hand-held photographic equipment.

The former, of course, are used by a nation's intelligence agencies. In contrast, the latter would probably be used on a rental basis by private investigators, industrial spies-for-hire, or investigative journalists.

History records that on June 26, 1794, the French Aerostatic Corps[28] used a balloon to make observations during the Battle of Fleurus. Given the name, *L'Entreprenant,* the observation balloon is reported to have stayed afloat for some nine hours, providing French military commanders with information about the Austrian Army's movements.[29] Communication with the ground

---

28. Which could be considered to be the first military air force.

29. More recently, it was reported that the private military contractor, Blackwater, had intended to develop surveillance blimps (Jeremy Scahill, *Blackwater: The Rise of the World's*

was via handwritten notes dropped from the balloon and by messages sent by semaphore.[30]

History also shows that during the First World War, the world's armies use aircraft not only for aerial combat, ground bombing and strafing, but for surveillance and reconnaissance. One such use was for spotting and ranging for artillery. These aircraft complemented observation balloons that were still being used as platforms for reporting enemy troop movements and directing cannon fire.

In the years between the French's use of surveillance balloons and the First World War, militaries developed the use of the skies as a means of gathering information. Italy used aircraft during the Italo-Turkish War (September 29, 1911, to October 18, 1912). History recorded that on October 23, 1911, *Capitano* Carlo Piazza flew over Turkish lines near Benghazi to conduct what would now be considered the world's first surveillance mission.

Captain Piazza later established another first by conducting the first photoreconnaissance flight in March 1912.[31] And, General Pershing used the U.S. 1st Aero

---

Most Powerful Mercenary Army (New York: Nation Books, 2007), p. 343.

30. Charles Coulston Gillispie, *Science and Polity in France: The Revolutionary and Napoleonic Years* (Princeton, NJ: Princeton University Press, 2004), pp. 73–373.

31. As an aside, the Italian military succeeded in conducting the first aerial bombardment on November 1, 1911. Second Lieutenant Giulio Gavotti is reported to have dropped three hand grenades on the Turkish-held Tagiura oasis, and another grenade on the Turk's bivouac at Ain Zara. Gavotti was also credited with flying the first night mission during the Libyan campaign on March 4, 1912. Walter J. Boyne, *The Influence of Air Power Upon History* (Gretna, LA: Pelican Publishing Co., 2003), p. 38.

Squadron to conduct reconnaissance during the Mexican Punitive Expedition, March 1916 to February 1917.

The Second World War saw an expansion of aerial surveillance in land and sea operations. Light aircraft were used for observation posts for artillery and mortar fire as well as observing enemy fleet from a distance. During the Vietnam War, observational helicopters were added to the variety of aircraft used.

During the Cold War, surveillance aircraft were used to spy on the Soviet Union, Cuba, and other unfriendly nations. Surveillance flights operated to ensure compliance with international treaties that, for instance, limited nuclear testing and weapons, in addition to other military concerns.[32]

Figure 10—U.S. Army Cayuse surveillance aircraft, the type used during the Vietnam War. Compliments of the U.S. Army.

---

32. Principally, to reduce the risk of another surprise attack like to one the Imperial Japanese Navy Air Service launched on December 7, 1942, at Peral Harbor.

The two most well-known spy aircraft are the Lockheed U-2 and the Lockheed SR-71. Each could carry a variety of payloads that included photographic equipment and signals intelligence sensors.[33] These aircraft could fly higher than other military aircraft, so they could conduct missions that were not possible with other planes.[34]

Until May 1, 1960, American U-2 spy planes penetrated deep inside the Soviet Union. When Francis Gary Powers was shot down by a surface to air missile, aerial surveillance made use of the newly developed SR-71 "Blackbird," which flew at three-times the speed of sound, so it could "outrun" surface to air missiles. But, the big step forward was the development of surveillance satellites.[35]

> On 16 March 1955, Headquarters Air Force issued General Operational Requirement No. 80, which officially ordered the development of an advanced reconnaissance satellite to provide continuous surveillance of 'preselected areas of the earth' in order 'to determine the status of a potential enemy's war-making capability.' That order officially put the Air Force in the space business, and the reason was reconnaissance.[36]

---

33. Signals intelligence (SIGINT) is the secret gathering of information by intercepting radio frequency signals that are transmitted person-to-person (COMINT) or via data stream from one electronic device to others (ELINT).

34. Such as the Boeing RB-47 that was used during the Cold War to collect electromagnetic intelligence on the Soviet Union. John M. Carroll, *Secrets of Electronic Espionage* (New York: E.P. Dutton & Co, 1966), pp. 174–176.

35. Also known as reconnaissance satellites, intelligence satellites, and spy satellites.

36. Mark Erickson, *Into the Unknown Together: The DoD, NASA, and Early Spaceflight* (Maxwell Air Force Base, AL: Air University Press, 2005), pp. 7–8.

Presently, satellites fulfill several military and national security surveillance roles: early warning; detection of nuclear explosions; signals intelligence; and optical and radar surveillance of targets on the earth.[37] In addition, there are many commercial surveillance satellites, and the high-quality images they produce can be purchased by civilians.[38]

From the point of view of gathering information for law enforcement, business competitor, and private party intelligence, satellite surveillance is available through several commercial outlets.[39] Law enforcement can also access government-controlled satellites if there is a pressing need, usually through a memorandum of understanding or legislation.

An example of how law enforcement agencies use satellite surveillance is to monitor remote regions of states or provinces that are difficult to access by vehicle or foot. The standout item being monitored is illegal drug crops, as well as the environmental impacts of these illicit crops.[40]

---

37. Robert M. Clark, *The Technical Collection of Intelligence* (Washington, DC: CQ Press, 2011).

38. In 2002, the director of CIA, George J. Tenet, "...ordered that imagery from commercial satellites become the 'primary source of data used for government mapping'' for the military and other agencies, and that the government's satellites be used only for such tasks in "exceptional circumstances." James Risen, "CIA Instructs Spy Agencies to Use More Commercial Satellite Photos," in *New York Times*, June 26, 2002, p. A11.

39. Harold Hough, *Satellite Imagery for the Masses* (Port Townsend, WA: Loompanics Unlimited, 2004), pp. 131–142.

40. Library of Congress Research Service, *Illicit Drug Flows and Seizures in the United States: What Do We [Not] Know?* (Library of Congress, 2019), pp. 1 and 2. Because satellite images use multiple sensors, such as infrared, analysts can distinguish different types of vegetation—e.g., cannabis, opium poppies, cocaine.

Hundreds of square miles can be monitored in a few hours that would otherwise take weeks to conduct by a search party on the ground. Because of this efficiency of satellite image interpretation, analysts can apply their skills to search for other hard to locate targets, such as fugitives hiding in forest areas.

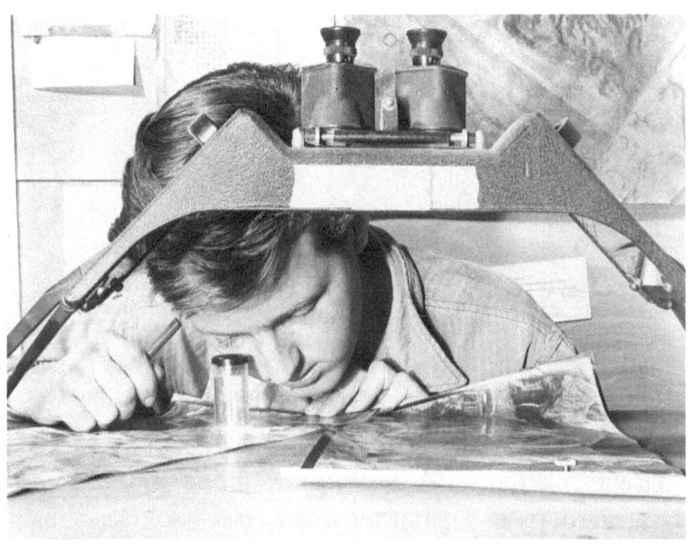

Figure 11—Analyst Sgt. James E. Kindseth, U.S. Fifth Air Force, Korea, identifies a Communist gun position, circa March 1952. Compliments of the Department of the Air Force.

Regulatory agencies also use satellite surveillance to monitor building construction in rural areas to ensure compliance with building codes and land tax purposes (i.e., new developmental work). Compliance agencies use satellite imagery to detect banned logging, illegal vegetation clearing, and unlawful water usage (e.g., taking water from rivers, streams, lakes, ponds, and the construction of prohibited artificial dams).[41]

---

41. Harold Hough, *Satellite Surveillance* (Port Townsend, WA: Loompanics Unlimited, 1991), pp. 189–98.

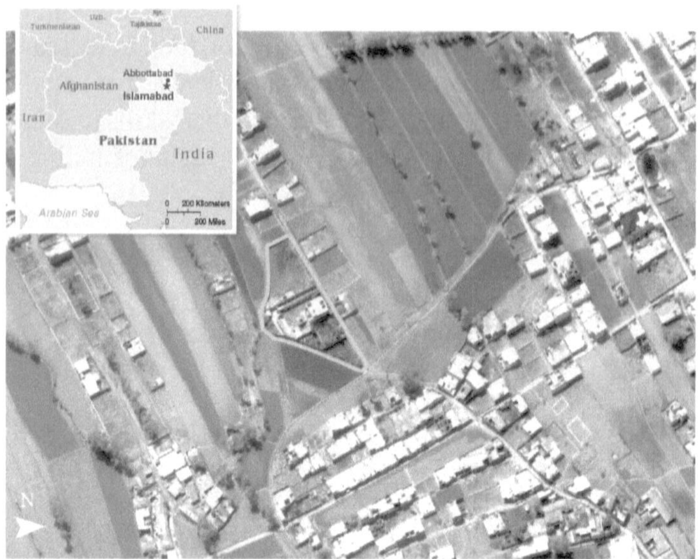

Figure 12—Surveillance satellite image of Osama bin Laden's compound, Abbottabad, Pakistan. Compliments of the CIA.

Although used by these enforcement agencies to monitor the environment, the same satellite resources can be used by private groups and environmental activists to establish a *prima facie* case for prosecution.

Private individuals have even used satellites to scrutinize areas of the world's seas and the lagoons around islands to help identify the likely resting places of sunken galleons containing treasury. Or, to search the world's jungles for crashed aircraft.

The business world has, in large part, abandoned the crude forms of industrial espionage used in decades past to embrace the use of open-source intelligence, including satellite surveillance. Businesses can now monitor their competitors' supply chain and logistics, manufacturing plants, mines, labor force issues (e.g., car park occupancy), or anything else by discreet overhead photographic surveillance.

Figure 13—A quadcopter surveillance drone, the type used by private investigators, activists, and in competitor intelligence.

If the area under surveillance is confined, obstructed, requires an oblique angle for clear viewing, or finer grain details are needed, then an alternative is to use drone surveillance.

These drones feature in many cinema and television productions, as well as having been shown in news footage and documentaries. Because of this coverage, we will not reiterate the particulars of these large pilotless aircraft, other than to state that their chief strength is that they can loiter many miles above a target, and silently observe what is happening on the ground.

What we will discuss is, instead, their use by the private and commercial sectors. Private investigators use drones to conduct spot checks on targets. Rather than conducting a "drive past" and risk being noticed, a drone can make observations from 125 yards above the target. Legal in many jurisdictions,[42] a drone "fly past" can take less than

42. This information is not intended as legal advice.

a minute and produce much information—arguable, more than a drive or walk past.

Because private investigators often want to know as much as possible about a target's dwelling and the surrounding area (i.e., background intelligence) before starting a surveillance mission, drones provide a convenient method. Take, for example, a stake-out. Some areas can be more alert to newcomers to the neighborhood, particularly in rural areas (which we will discuss in the next chapter). Therefore, drones offer a high-level utility.

Private investigators (and political activists[43]) use these machines to gather intelligence and evidence. The quality of still images and video from the onboard cameras is usually good enough to be accepted in evidence in courts of law. These images can be comparable to photographs taken on a smartphone. Yet, the operator remains at a safe distance, and the filming can be discreet, avoiding tipping the hand of the surveillant.

---

43. Henry Prunckun, "Contre-Espionnage: Tout ce qui est Ancien est Nouveau," in Paul Charon, Jean-Baptiste Jeangène Vilmer (dir.), *Le Renseignement: Approches, Acteurs, Enjeux*, (Paris: Presses Universitaires de France, 2022).

— CHAPTER SIX —

RURAL SURVEILLANCE

When discussing rural surveillance, we are referring to making observations of either a person or a premise in a lightly populated area. A rural setting often means that operatives need to enter the area, remain there, and exit covertly.

Some rural settings are more challenging than others, particularly open countryside. Without the cover afforded by woodland, scrub, or other obscuring terrain features, open countryside presents a hazard to the surveillant's personal safety by exposing them to the target, and if this happens, compromising the mission. For this reason, operatives who undertake this form of surveillance are trained specifically for this type of work.

Training for rural surveillance usually includes courses in fieldcraft, map reading and navigation, tactical movement, cover and camouflage, tracking, survival, and first-aid. Specialized courses in rural surveillance usually assume knowledge of surveillance methods, although a refresher on how certain pieces of surveillance equipment are used in a rural setting can also form part of these courses. For instance, there are heightened risks of being spotted by the glare from binocular and camera lenses.

Because this book is not focused on these types of fieldcraft, we will defer discussion to texts that address these subjects directly.[44]

---

44. Van Ritch, *Rural Surveillance: A Cop's Guide to Gathering Evidence in Remote Areas* (Boulder, CO: Paladin Press, 2003).

— CHAPTER SEVEN —

TECHNICAL SURVEILLANCE

Technical surveillance is a broad term that covers several types of technology-aided observation. In this chapter we will discuss optical surveillance, covert photography, electronic audio surveillance, Internet surveillance, GPS tracking, and postal surveillance.

OPTICAL SURVEILLANCE

Because surveillance relies almost entirely on their eyesight, various devices are used to assist in extending their vision. This technology also helps surveillants to position themselves further away from the target, thus reducing the possibility of discovery.

The principal device used in surveillance is a pair of binoculars because they are light and portable. Binoculars are optical devices consisting of two prism-operated telescopes fixed in parallel. This configuration enables the surveillant to view a magnified image of the target using both eyes, thus rendering a wider view than a monocular.[45] Binoculars have been designed to provide both magnifying power and light gathering capabilities. The latter is essential for surveillance work at dusk and night.

When viewing a target at distances that exceed the effective range of binoculars, another optical device is used—a telescope. The magnifying power of a telescope ranges from twenty to several hundred times that of normal vision. Binoculars, by contrast, range from about six to

---

45. A monocular is well suited to situations where a surveillant only needs to briefly view the target.

twenty times that of normal vision. A monocular is often less than binoculars.

A viewing device usually associated with submarines but is frequently used in land-based information gathering is the periscope. Small portable high-quality devices are used to peer into high, inaccessible windows, over walls and around corners. Periscopes are also used in applications such as surveillance vehicles. Versions of the periscope concept are, for instance, hand-held digital cameras raised over obstacles by improvised implements, such as poles, thus functioning as a periscope.[46]

Low light (night) viewers are devices that are either "active" or "passive." The first group operates by using an infrared light beam. The surveillant projects the invisible beam of energy so that it illuminates the target or area under watch. The image is then viewed with special equipment, which converts the infrared radiation into the visible light spectrum. The second range of devices operate by amplifying the existing background light—moon, stars, streetlights, and so on—by several thousand times, thus literally turning night into day through the sights of the surveillant's night scope.

COVERT PHOTOGRAPHY

Arguably, the most used camera in covert photography is the digital single-lens reflex (D-SLR). However, in some unique situations, such as surreptitious copying of documents, a subminiature camera might be used. Obviously, the value of a subminiature camera lies in its concealable size. The advantage of a D-SLR is the choice

46. Henry W. Prunckun, *Information Security: A Practical Handbook on Business Counterintelligence* (South Australia: Bibliologica Press, 2020), pp. 13–14.

of either "fast" (light-sensitive), wide-angle or telephoto lenses that render high-quality images.

Telephoto lenses play an essential role in surveillance photography. Reflective mirror (catadioptric) lenses not only enable surveillants to significantly reduce the physical size of their equipment, making concealment easier; they act to extend the operative's operational distances. Catadioptric lenses use a system of mirrors to compress the light's optical path. Magnification (measured in the lens's focal length) varies from about one hundred to two thousand millimetres.

Digital video cameras are also used for surveillance work. These cameras are no larger than a D-SLR, and often have a reasonably effective zoom telephoto lens, thus allowing surveillants to record their observations from a safe distance. In the case of static surveillance (a stake-out), a "time lapse" option can be used to provide several days of continuous coverage before changing the memory card.[47] Time-lapse photography operates in a single frame mode at greatly reduced speeds. Other possibilities for static surveillance can be imaginative, like cameras built into briefcases, books, wall clocks, artworks, ceiling-mounted smoke alarms, and decorative indoor plants.

ELECTRONIC AUDIO SURVEILLANCE

The use of electronic devices in surveillance has become a pervasive and permanent phenomenon. The rapid advance in the state-of-the-art of electronics and the open accessibility of this technology has placed surveillance devices within reach of everyone. Sophisticated devices can be purchased over the Internet for around $100. Yet, intelligence and law enforcement agencies and the military

---

47. However, there may be cases where the video could be live streamed, which would be superior. But, is some "denied areas" a time-lapse option could be a realistic alternative.

have access to the most advanced electronic surveillance devices for surveillance.

Nonetheless, anyone can purchase electronic components and, if they have a high school understanding of electronics, can construct a bug, wiretap, or another listening device. Over the past decades, there have been many quasi-underground publications that provide schematics and instructions.[48] Even *Popular Mechanics* has presented articles on surveillance devices that, during the Cold War, CIA operatives recommended as "...less costly than similar device[s] in [the CIA's] inventory."[49]

The fundamental principle of any audio surveillance operation is to be able to plant a quality microphone as near as possible to the target(s) and, in doing so, avoid background noise that may make the intercept unusable. The types of microphones used in audio surveillance are dictated by the nature of the work. In general, they need to be very small, referred to as *sub-miniature*.

Their minute size allows them to be secretly deployed in the environment under surveillance. Once in place, they can be connected to a high-gain amplifier, and then to several types of output devices: headphones (for live monitoring); a digital recorder (for listening later, or for intelligence and evidentiary purposes); a transmitter to be streamed over a wireless network (e.g., Wi-Fi, Bluetooth, and/or the Internet).

---

48. By way of example, see William Powell, *The Anarchist Cookbook* (New York: Lyle Stuart, 1971), and *Portfolio of Schematic Diagrams for Electronic Surveillance Devices* (Flushing, NY: Mentor Publications, 1979).

49. Jack Devine with Vernon Loeb, *Good Hunting: An American Spymaster's Story* (New York: Sarah Crichton Books, 2014), p. 52.

There are several types of microphones the surveillance operatives can use depending on the situation. They are the tube, contact/spike, pneumatic, and directional microphones.[50]

Tube microphones are designed to be inserted into a targeted room via a tiny drill hole. This type of microphone consists of an element connected to a thin tube. The tube emerges flush with the wall in the targeted room and, if installed correctly, can only be detected by close inspection. Tube microphones are likely to be in spots made classic in spy novels—behind a wall, a picture, a piece of furniture, that is anything that would hinder detection for as long as possible.

Contact and spike microphones require less effort to secure their installation. These microphones do not respond to air vibration as a conventional microphone does but instead translates vibration into sound. Contact microphones are like the "pick-ups" used by musicians to amplify their instruments; spike microphones are akin to the phonograph needle used to play vinal records. These types of microphones are simply attached to the exterior of a wall, window, floor, or ceiling. Once in place, these devices will reproduce the sounds made within the targeted room. These microphones lend themselves to permanent installation.

The pneumatic cavity microphone is the electronic version of the glass tumbler against the wall trick, historically recognized as an effective method for monitoring adjacent room conversations. This type of microphone is superior because it operates by using a specially constructed shell responsive to surface vibrations at audio frequencies found in the range of human speech.

---

50. Prunckun, *Information Security: A Practical Handbook on Business Counterintelligence*, 2020, pp. 17–18.

This cavity is used in conjunction with a microphone element to enhance the microphone's performance. It also forces audio output to correspond to a wall's surface (or window, floor, or ceiling depending on the case) vibrations rather than a direct sound output.

Directional microphones are of two types: parabolic and shotgun. Parabolic microphones consist of a "dish" with an inwardly pointing microphone element. The targeted audio is reflected from the dish to be focused on the microphone element, thus gaining a directional effect.

The shotgun (also known as a rifle or machine gun) microphone operates on the same principle but uses a long tube, or set of tubes, in a cluster to home in on the targeted conversation. The effective range of these microphones is reported to be about 150 meters, and they are known to be able to pick up audio through closed windows at closer distances.

Another area of electronic audio surveillance is that of wireless microphones, also known as miniature transmitters. These devices rank further up the hierarchy in surveillance sophistication.[51]

Transmitters offer an agent much greater safety from detection because installation is virtually effortless. These devices do not have any wires or connections to reveal their location. They can be attached to furniture or office/household fixtures by means of a magnet or adhesive surface or concealed in everyday objects such as rings, pens, books, pens, or photo frames.

Transmitters do not require surveillants or their equipment to be located nearby. Their transmission range varies from sixty yards to half a mile. It is directly

---

51. Prunckun, *Information Security: A Practical Handbook on Business Counterintelligence*, 2020, pp. 18–20.

dependent upon the transmitter's output power, the thickness of surrounding walls, and the sensitivity of the surveillant's receiver.

Body transmitters are generally larger, more powerful, and better constructed than wireless microphones. This is because the chance of detection is lower, and the device is intended to be used repeatedly. These devices are designed to operate from a surveillant's coat pocket, underclothing, or attached by tape directly to the operative's body.

Akin to the body transmitter is the briefcase or laptop bag transmitter. Not only can this device be used by a "walk-in operative," but it can be conveniently "forgotten" in an office to obtain the ensuing conversations.

Another type of electronic surveillance transmitter operates by broadcasting in the very low frequency (VLF) range. This device uses electrical power lines for signal transmission. The signals move along the AC wire path, and because of the device's very low frequency, little energy is radiated into space. This method of communication, although less used than it once was, is used by many of the so-called wireless intercoms.

The telephone is arguably *the* medium of electronic surveillance. Audio "penetrations" involve two methods. The first method uses legacy landline devices where conversations are intercepted directly from telephone lines.[52] As such, there is no requirement to enter the target's premises. The second method employed in landline systems uses a portion of the telephone system for room surveillance but does require physical entry into the premises.

Because the telephone company provides the electrical power required to operate a subscriber's landline service,

---

52. Scott French, *The Big Brother Book* (Secaucus, NJ: Lyle Stuart, 1975), p. 48.

this medium offers some benefits for surveillants. For example, the telephone's power supply can be used to operate electronic audio devices, the wires themselves can be used to carry the resulting audio signals, and the microphone in the handset can be used to listen in on room conversations (e.g., hotel rooms).

The techniques used to tap conversation from landline telephones consist of (a) direct wire connections and (b) induction coils. With a direct wire connection, the line is cut, and the listening device spliced in place. With induction coils, the tapping process literally lifts the audio signals off the telephone line by inductance, thus requiring no splicing. For this reason, induction coils may prove to be undetectable by electronic countermeasure sweeps. The only effective way to detect this type of bug may be by visual inspection of the telephone's wiring.

The placement of landline telephone surveillance devices can vary; the only limit to their deployment would be the ability for the operative to gain access to the target's telephone system. The device may be installed within the telephone instrument itself, anywhere along the line in the targeted building, on a telephone pole/pit outside the building, or in the wiring closet or terminal room where the lines are joined to the branch feeder cable.

The ubiquity of cellular telephones has permitted surveillants to eavesdrop on voice communications, but this medium also allows them access to every piece of information stored on the device or that passes through or is generated by the phone.[53]

Almost every Hollywood spy thriller features the protagonist surreptitiously accessing the other's cellular data. The scenarios are predictable—invariably, a spyware

53. Kevin D. Murray, *Is My Cell Phone Bugged?* (Austin, TX: Emerald Book Company, 2011).

app is installed on the phone, or the phone is hacked and taken over by the perpetrator. Then the information is "milked" from the device or identified and tracked the target's whereabouts using the hand-held's GPS function.

Think for a minute what might be at stake: the target's list of contacts; the ability of the hacker to send false messages or make calls; access voice mail messages and phone logs, and of course, receive alerts when the target is making or receiving calls or activate the device's microphone remotely.

Are these situations the storyteller's fertile imagination manifesting itself on the pages of novels or the widescreen? No, these are examples—though dramatized for their audiences—of what technical surveillance is capable.

Software applications and surveillance hardware used to "infect" a target's device are so numerous that even a summary of them is beyond the scope of this book.[54] Suffice to say that if we look at the security advice in the public domain about avoiding cell phone surveillance, we can understand how penetration is done and how to prevent this happening.

The fundamental tenet is, do not have the device in the vicinity of a sensitive conversation. Recall that when entering a secure area, such as an embassy, computer data center, research laboratories, or law enforcement or military facility, etc., mobile devices are left at the entrance. What does this suggest? That the devices are not to be trusted. These potential targets treat cellular devices as if they are all compromised. Therefore, one could conclude that if this is the view of such targets, it is easy to compromise a device. And this would not be an unreasonable inference.

---

54. See, for example, Greg Hauser, *Pretext Manual* (Austin, TX: Thomas Investigative Publications, 1994).

Think about how fiction portrays scenes where people meet to discuss important information. Not only do they practice the surveillance detection methods discussed in Chapters 8 and 9 on anti- and countersurveillance, but they either leave their cell phone at home or the office, or they remove the SIM card *and* the battery.

When personnel employed by these potential targets do need to use a smart device, they use secure devices. These devices are specifically designed to guard against technical surveillance. These provide the most robust protection against penetration by hostile software infection and interception to Top Secret.[55]

Another dimension to electronic audio surveillance is the system that operates using directional beams of light energy. These systems are based on the use of lasers of either visible or infrared energy to convey the intercepted audio. They are reported to be reasonably reliable and virtually undetectable.

These systems consist of a laser light source that focuses its energy on a window in a targeted room and an optical receiving/decoding device. The system operates by detecting minute vibrations of the reflected radiation. The decoding of these vibrations reproduces the conversations or sounds within.

---

55. Information is classified according to the potential harm unauthorized disclosure could cause. Various nations (as well as corporations) have definitions that suit their needs; but as an example, levels may resemble these: Unclassified; Official; Restricted; Confidential; Secret; and Top Secret. A classification of *Secret* would be defined as information that could cause "serious damage" to national security. *Top Secret* is assigned to a nation's most sensitive information—information if made public would cause "exceptionally grave damage" to national security.

Interception of computer data is accomplished by intercepting the target's computer's in-bound and out-bound data via the Internet. Therefore, Internet surveillance includes monitoring computers, which incorporates servers, as well as the electronic networks that connected them.

This covert monitoring can be initiated by governments (e.g., law enforcement and intelligence agencies), spies-for-hire (i.e., illegal industrial espionage), organized criminals (say, for the purpose of ransom attacks or outright theft), or individuals (for many reasons, ranging from curiosity, domination,[56] and psychosis to hostile nation-state espionage).

Internet surveillance by governments is necessary because it allows agencies to monitor criminal threats to public order and safety and prevent crime and investigate illegal activities. Take, for instance, the ransomware attack on Colonial Pipeline in May 2021. The computer system that managed the company's oil pipeline, which ran from Texas to the southeastern states, was shut down by a virus that prevented the system from operating.

This example was not a trivial matter because the pipeline conveyed essential supplies of gasoline and aviation fuel. The ransomware attack resulted in shortages at approximately 71% of filling stations, which caused panic buying.

It is hard to argue against an agency like the FBI having Internet surveillance capability because it was central to

---

56. For instance, keystroke loggers installed on a personal computer by a jealous or controlling spouses/partners.

prosecuting the criminal gang that executed the attack.[57] Although there are critics of government surveillance, we need to remember that any proposal to undertake monitoring is first subject to a court-approved warrant. Yet, when Internet surveillance is conducted by a less than benevolent state, the impact on privacy and the free exchange of thoughts and ideas can become Orwellian. As an illustration, take the situation in China. The Chinese Communist Party maintains a network of systems that monitors its citizens as part of its Social Credit System.[58]

Internet surveillance is just one of several systems that comprise the Chinese Communist Party's mass surveillance program. With over 750 million Internet users in China, the government has implemented elaborate electronic censorship and propaganda mechanisms to control the country's population.

"Civil liberties activists in the West keep their governments honest. In China, they are thrown in jail."[59] Yet, there are anti-surveillance measures that some Chinese citizens use to circumvent government control. These tactics will be discussed in Chapter Eight.

---

57. It was alleged that the *DarkSide* cyber-criminal hacking group was responsible. It is understood that this group was an Eastern Europe-based gang that extorted their victims, such as the Colonial Pipeline attack. Dustin Volz, "U.S. Blames Criminal Group in Colonial Pipeline Hack," *The Wall Street Journal*, May 10, 2021.

58. This is essentially a blacklisting process for those who do not comply with the government's policies. It is based on a centralized record system that assess each person or business's "trustworthiness." For those who comply, they are whitelisted and can gain privilege, services, and mobility.

59. Clive Hamilton, *Silent Invasion* (Melbourne: Hardie Grant Books, 2018), p. 251.

Knowing where a target is, where they have been, and the time spent at different locations can be important information for either evidence tendered in court or for intelligence purposes. Collecting these data can be via surveillance operatives where more subtle observation may need to be required—as in the case of servicing a dead drop—but an arms-length approach could be facilitated by a Global Positioning System (GPS) tracker.[60]

Geo-tracking units are devices that can be secretly attached to vehicles, boats, cargo, goods, people, or other targets. Using, say, a cellular network, these devices transmit the target's geographic position.

At the heart of these devices is a radio that receives GPS satellite signals how a car GPS unit does and transmits the target's coordinates to the surveillant. Trackers can be categorized as either 1) data loggers, 2) data pushers, or 3) beacons. These trackers can be used as-is from the manufacturer or built into everyday objects, thus making them covert.

Loggers record the target's position at preset intervals that are stored in the device's internal memory. This memory can vary, from a built-in memory card to an externally attached USB flash drive. In any case, the purpose is not to transmit the data in real-time but to record the target's movements so that when the device is retrieved, the information can be downloaded for analysis.

---

60. A GPS tracker could be used in the preliminary stages of an investigation to establish the target's route(s) and patterns. In this way, physical surveillance can be setup to observe certain parts of the route, thus reducing the target's ability to detect a tail.

Data pushers operate the same way as a logger but also have a transmitter that sends these data via a cellular network connection to a receiving computer. A satellite-based tracker will operate anywhere on the globe. Data pushers are, arguably, the most prevalent type of GPS trackers. Cell phones are capable of operating in this mode even when the phone is turned off because it stores the geocoordinates for transmission when the phone is switched on again.[61]

GPS beacons send the device's position along with data like speed or altitude to the surveillant's computer. Beacons are used where the target's location needs to be checked occasionally. Shipping containers that contain contraband presents as an immediate example.

POSTAL MAIL COVER

In an age of ubiquitous computing, information garnered from postal mail may not present as a viable surveillant technique. Yet, items transported by a nation's postal service can yield valuable information.[62]

Law enforcement agencies use this as an investigative technique to monitor targets. The process involves the recording of the information contained on the *outside* of the envelop or package as they pass through the postal system before being delivered.

In the United States, the US Postal Service records these data and sends it to the requesting agency. Mail covers are used for investigating criminal activity, locating a fugitive from justice, identifying items

---

61. On cell phones manufactured after 2014, see *Find My Phone* feature.

62. Alternative Technologies Information Service, *The Postal Mail Cover* (Buffalo, NY: Alternative Technologies Information Service, 1984).

forfeitable under law, or cases that involve national security. So, it is not a method that can be sued by industrial spies, private investigators, or paparazzi. This is because it requires authorization by the Chief Postal Inspector (or their delegate). Even though only the exterior details are recorded, U.S. law mandates a written request based on "reasonable grounds" that must be stated in the application.[63]

As an indicator of how often this surveillance technique is used in the United States, USPS statistics that the number of mail covers increased from 9,022 in 1984 to 14,077 in 2000.[64]

Nonetheless, since 2001, the USPS has, in effect, conducted mail covers on all U.S.-based recipients of postal items under the authority of the Mail Isolation Control and Tracking program.[65] This program was initiated in the wake of the 2001 anthrax letter attacks that killed five people, including two postal workers.[66] Given this method's value to law enforcement in keeping its population safe, it is inconceivable that other nations do not have an equivalent type of postal surveillance capability.

---

63. U.S. Postal Inspection Service, *Mail Cover Requests Publication 55* (Washington, DC: U.S. Postal Inspection Service, 2006).

64. Elizabeth Amon and Michael Ravnitzky, "'We're Getting Popular,' Says a Mail Cover Supervisor," in *The National Law Journal*, April 1, 2002.

65. U.S. Postal Service, *Mail Isolation, Control, and Tracking: Management Alert Report* (Washington, DC: Office of Inspector General, 2013).

66. Ron Nixon, "U.S. Postal Service Logging All Mail for Law Enforcement," in *The New York Times*, July 4, 2013, Section A, p.1.

— CHAPTER EIGHT —

ANTI-SURVEILLANCE

There are two methods to avoiding surveillance—defensive and offensive. Or, in the terminology of the trade, anti-surveillance and countersurveillance.[67] In this chapter, we will discuss defensive countermeasures. Offensive measures will be examined in Chapter Nine.

PASSIVE PROTECTION

Anti-surveillance is a group of passive protective measures that comprise several tactics that, when used alone or in combination, reduce the risk of being surveilled. We will assess four countermeasures—electronic, software, structural, and human.

Electronic countermeasures cover a wide range of "bug" hunting methods and devices. Known as a *sweep*, technicians use sensitive radio frequency receivers to survey an area looking for electronic (or magnetic) emissions that could indicate a hostile listening device. Such a sweep is used in conjunction with a physical search

---

67. Although this terminology is not used consistently by some scholars. For, instance, ACM IV Security Services used the terms in reverse; that is, anti-surveillance to mean active (offensive) measures and countersurveillance to mean passive (defensive) measures; see, ACM IV Security Services, *Countering Hostile Surveillance* (Boulder, CO: Paladin Press, 2008), pp. 17–21. Yet, in another publication, this same author used it correctly; see, ACM IV Security Services, *Surveillance Countermeasures* (Boulder, CO: Paladin Press, 2005), pp. 127–132.

of the area because some listening devices can be switched off to help guard against detection.[68]

If media reports of the number of cases involving cyber-surveillance are even half as extensive as one is led to believe, then the phenomenon is pervasive. A technical countermeasures sweep of a computer system is akin to the type conducted in the physical world—an electronic probe throughout the system and a physical examination of all computer hardware and network cabling, routers, switches, and modems.

Structural countermeasures are about securing the physical environment in which people and the apparatuses they use work. Starting with perimeter controls—fences, walls, and the like—to security clearances and access control to these buildings. It also includes access to and the handling of information within the organization's building(s).

The list of processes and procedures that govern the application of structural countermeasures is wide-ranging, but touching on a few that will give us some scope of the depth and breadth of the strategies consider the following: locks and barriers, alarm systems, safes, preventive acoustics for windows, electromagnetic shielding for electronic equipment and cabling, prohibition of personal electric devices in the work area, and sweeps at unpredictable intervals.[69]

---

68. Or they are passive devices that are activated by directing a beam of radio frequency energy at them, like the device hidden in the Great Seal presented to the then- Soviet Ambassador W. Averell Harriman in 1945.

69. For a detailed discussion of these types of countermeasures, see Hank Prunckun, *Counterintelligence Theory and Practice, Second Edition* (Lanham, MD: Rowman & Littlefield, 2019).

If we consider a maximum-security prison, we are unlikely to be faulted in our reasoning if we were to think that the strategies these detention facilities used to keep contraband out are state-of-the-art. Yet, news reports show that drugs, guns, cell phones, and other prohibited items regularly find their way into jails. Why? Because the weakest link in the security chain is the human element.

Likewise, whether it is complacency, corruption, or human error, people are likely to be the vector by which an espionage agent can get a device into a building targeted for monitoring. Or, an operative can obtain information by using a pretext to manipulate an employee to allow entry.[70]

Therefore, it follows that any strategy that limits the ability of people outside an organization to leverage the goodwill, laziness, or naivety of employees will reduce the risk of surveillance. The cornerstones of this approach are being situationally aware, avoiding risky personal encounters with strangers, and keeping a low personal profile.

STATIC AND MOBILE SURVEILLANCE

Intuitively, it might seem that it would be more challenging to detect mobile surveillance than fixed surveillance, but this may not be so. How can this be, when a surveillant is on the move, they can use disguises and coordinate their movements by concealed communication devices? In contrast, fixed surveillance

---

70. Ahern describes numerous cases involving private investigators who have penetrated security conscious organizations by exploiting people's weaknesses. See, Frank M. Ahern with Eileen C. Horan, *How to Disappear* (Guilford, CT: Lyons Press, 2010).

would appear easier to spot, for instance, someone sitting in a car or standing on a street corner.

The latter situation paints a common—or, perhaps a ubiquitous—picture of how static surveillance is conducted. Yet, it overlooks how more sophisticated observation posts are established. To demonstrate this, take the example of where a member of an outlaw motorcycle gang is being surveilled. If they are on the watch for surveillance, they may look out their apartment's window(s) for people in vehicles. They may look for people milling about, sitting on benches, standing in shop doorways, or other "waiting" behavior types.

Figure 14–Watching for a vehicle pulling away from the curb.

If they do not see anyone, they may then leave the building, and as they do, glance left, right, and across the street for signs of a tail. They may walk left, get to the corner, and then turn back, walking in the opposite direction to see if a vehicle has pulled away from the curb or if anyone has appeared from a concealed location.

If they detect no signs were detected, they might think they are in the clear. However, what they are not able to see is that there is a surveillant in a window on a sixth-floor apartment building three blocks away, watching the gang member's doorway as well as the entire townhouse.

Depending on the vantage point, distant observation posts like this example can radio mobile units further ahead to pick up the tail. So, detecting static surveillance is not as easy as looking around for some clumsy "gumshoe."

In terms of anti-surveillance measures, one needs to assume that any viewing position visible from a given place can be used in reverse—to spy on those in that location.

Mobile surveillance can work in conjunction with static surveillance or be used by itself. Therefore, anti-surveillance measures need to consider various scenarios; nevertheless, moving surveillance is more straightforward to detect than static surveillance. This is because the surveillants need to follow the target.

The target chooses where to go, how fast or slow to travel, when to make stops or not, their mode of transport, and if and when to change modes. The target causes the surveillant to respond to them, therefore has the advantage in that they can assess travel time, distance, and direction,[71] three factors that we will discuss in the next chapter.

---

71. Michael Chesbro, *The Privacy Handbook* (Boulder, CO: Paladin Press, 2002), pp. 182–188.

— CHAPTER NINE —

COUNTERSURVEILLANCE

We have all seen a scene in espionage movies where an operative leaves for a meeting but engages in a series of moves to see if anyone is following. They may drive around a traffic circle twice, do an unexpected U-turn on a busy road, cross a bridge, or park and proceed on foot. These moves are part of what forms offensive security known as *countersurveillance*.

The consequences of not checking for a tail can be, on the surface, benign. This is because an opponent may be seeking information about the target's associates, schedules, and other factual matters. While being followed, the target is unaffected, but later, the target's business or its associates may suffer irritating repercussions.

Figure 15—The late-Richard Welch. Photograph compliments of the Central Intelligence Agency.

At the other end of the scale, it could mean death. Take the case of Richard Welch, who was a career CIA officer. He was the Chief of Station in Athens, Greece, when, in 1975, he was assassinated by the radical left-wing Revolutionary Organization 17 November.

His assassination underscores several countersurveillance issues that he should have practised, but in the context of this book, his failure to identify surveillance led to him being able to be followed by four men in a stolen vehicle on his way home. He was then approached without warning and killed at a close range.

When countersurveillance is executed, it can save lives. Take the case of J. Cofer Black,[72] a thirty-year veteran of the CIA and later, a leading figure in the private security firm, Blackwater. In 1994 while posted by the CIA to Khartoum under cover of a minor U.S. embassy diplomat, Black was involved in monitoring Osama bin Laden. While Black was directing his CIA assets to keep bin Laden under surveillance, bin Laden was keeping Black under surveillance. When this happened,

> Black and his case officers picked up this surveillance and started to watch those who were watching them. ... The CIA officers saw that bin Laden's men were setting up a 'kill zone' near the U.S. embassy. They couldn't tell whether the attack was going to be a kidnapping, a car bombing, or an ambush with assault rifles, but they were able to watch bin Laden's group practice the operation on a Khartoum street. As the weeks passed, the surveillance and countersurveillance grew more and

72. Joseph Cofer Black served the CIA with distinction in several roles within the Directorate of Operations, the part of the agency that was responsible for carrying out covert and clandestine operations overseas. Jeremy Scahill, *Blackwater: The Rise of the World's Most Powerful Mercenary Army* (New York: Nation Books, 2007).

> more intense. On one occasion, they found themselves in a high-speed chase. On another, the CIA officers levelled loaded shotguns at the Arabs who were following them. … Exposed, the plotters retreated.[73]

⌘ ⌘ ⌘

As we discussed in Chapter Eight, there are defensive (passive) measures that can be used to foil a surveillance operation. Countersurveillance is the offensive form of the craft (active measures). *Countersurveillance* is an intelligence term for determining if you are being followed. This is an important skill because if you are being tailed, then you can decide whether what you are doing is going to reveal information that might help the opposition, compromise others, or result in you being kidnapped or killed.

> Since surveillance is always possible, [counter]-surveillance can be employed even when there is no specific indication that surveillance is present. In fact, professionals in covert activities, such as espionage agents, invariably employ [counter]-surveillance activities as a standard practice, due to their extreme need to ensure that their operational activities remain unobserved and undetected.[74]

At the core of countersurveillance is detection. How you detect surveillance is more of an art than a science, yet like the fine arts, some fundamentals guide practice. The first principle is guarding against complacency. This does not mean you need to become a nervous neurotic; it means

---

73. Steve Coll, *Ghost Wars: The Secret History of the CIA, Afghanistan, Bin Laden, from the Soviet Invasion to September 10, 2001* (New York: William Morrow, 2004), p. 271.

74. ACM IV Security Services, *Countering Hostile Surveillance*, p. 20.

staying situational aware. Situational awareness is a topic taught at many security and intelligence agencies, such as at CIA's clandestine training facility, colloquially known as *The Farm.*[75]

At this high threat level, clandestine intelligence officers and their agents must practice a corresponding acute level of surveillance detection. In other spheres—such as business—the detection techniques correspond to that level of threat, which is manifested by the types of surveillance an opponent is practising. In this chapter, we will discuss surveillance detection in the context of mobile and foot surveillance, but not static, aerial, or technical surveillance.

Surveillance detection rests on situational awareness, but how is that put into practice? There are three techniques that a situationally aware person uses.

The first method is the distance travelled assessment. As its name implies, a potential target assesses whether people or vehicles behind them have been there for longer than expected.

This is, however, a relative measure. Suppose a trip from a business's head office to a warehouse or manufacturing facility is only several kilometres away. In that case, it may not be unusual for other vehicles to travel the same route. It is, after all, only a short distance. Likewise, if a potential target is travelling across the city on a freeway, no doubt a lot of vehicles will be doing the

---

75. Officially known as *Camp Peary.* It is a 10,000-acre military reservation in Williamsburg, Virginia under the legal authority of the Department of Defense. Tom Mangold, *Cold Warrior: James Jesus Angleton—The CIA's Master Spy Hunter* (New York: Simon & Schuster, 1991), p. 183.

same, even getting on and off the highway at the same entrance and exit ramps.

The second method is time travelled assessment. Perhaps a better gauge because the longer a potential target travels, the more likely any tail will be noticed. Even if the opposition is a well-funded security or intelligence agency and can have afforded to task several teams to rotate in-and-out of the pursuit, it is unlikely that they will have unlimited vehicles and operatives.

The odds are that those following be able to have enough vehicles to change as the tail progresses. At some stage, situational awareness will alert the target to the fact that they saw that vehicle before. This could mean, on the same day or during previous days. The same applies to surveillants on foot.

This method may also produce false positives. If the target is driving around a city for an hour or two trying to detect surveillance, they may encounter another vehicle(s) coming and going too. The target may mistakenly take this coincidental sighting to mean they are being followed.

The third method may be the most telling. That is, a change in direction. If a target changes their direction of travel, whether they are on foot, bicycle, or in a vehicle, it immediately confounds the surveillant. In such a case, the tail cannot do the same; otherwise, they will show their hand. So, they must break contact or pass on the role of "the eye"[76] to another team.

As good as this method is, it may also produce false positives, but is less likely, nevertheless, if the three assessments are used together. This is what is termed a

---

76. Whoever is observing the target is referred to as "the eye." T.J. Waters, *Class 11: Inside the CIA's Post-9/11 Spy Class* (New York: Dutton, 2006), p. 71.

surveillance detection route (SDR). An SDR can be used whether a target is on foot, bicycle, or in a vehicle.

What can a target do if they discover they are being monitored? Their first thought is likely to be to protect their operation, whether that is an agent meeting, servicing a dead drop, or collecting intelligence (e.g., covert photography). This does not mean that they will re-enact a scene from a James Bond or Jason Bourne film by trying to lose the tail (i.e., aggressive escape and evasion driving techniques categorized as *counter-pursuit*); they will simply abandon the operation.

Because the target is in control, they can slow the pace of the hunt, speed it up, or thwart the effectiveness of those tailing them. Take, as an example, a target might hurry into a café to order a coffee. As they sip, they read one of the complimentary newspapers. When they decide that they have wasted the surveillant's time, they place a piece of paper into the newspaper and lodge it, not back on the display rack, but in an unusual place in the café.

At this point, the surveillance team needs to decide whether they stay and monitor the café to see who collects the newspaper or maintain the tail. If this situation was not anticipated in the team's SMEAC briefing, they need to radio or telephone to get guidance, all the while the target can be moving on to conduct another impromptu deception.[77]

---

77. Opportunities for performing improvised tricks relies on the operative's imagination. A few examples include standing next to someone to fake passing off information, shaking hands with a stranger, patting a dog (especially around its collar where a message could be hidden) while chatting to the owner, and so on. In all these cases, these people could be a legitimate (random) contact, but the only way to establish this is to follow each of them.

What we have discussed so far is where the target of the surveillance operation (i.e., sometimes referred to as the *principal*) enacts countermeasures themselves. A more sophisticated method involves another operative or a team of investigators that conduct surveillance of those who are surveilling the target.

Countersurveillance operations usually follow a planned surveillance detection route that features one or more surveillance detection points (SDP). Surveillance detection points are important sites because they force the opposition's surveillants into a position where they can be observed. If there are several of these SDPs along the route, it affords the ability for more sightings of hostile surveillance, and this, in turn, adds certainty to any conclusion of being tailed.

Countersurveillance may present as surveillance, and it is, but the task calls for far more sophisticated tradecraft. Take as an example, disguises. A surveillant many don a disguise part way through a follow to camouflage their appearance. In a countersurveillance operation it is likely to happen in reverse—starting in disguise and shedding the camouflage as the tail progresses. This is because taking off a costume is easier and faster than putting it on.[78]

Intra-operational communication is another aspect where countersurveillance differs. Yes, each countersurveillance operative will use a concealed communications device, but the principal will not. The reason for this is apparent—the principal is under hostile surveillance, so their every move, motion, or action is

---

78. Items can also be left behind rather than carried in bags or backpacks to be put on during the operation.

under observation. Any attempt to speak into a hidden cuff mic, for instance, will indeed be seen.[79]

More thoughtful signalling is required. For instance, a signal needs to be more than a hand motion, such as rubbing one's hand through their hair or scratching one's head. This is because these types of movements take place quickly, and, in all likelihood, the countersurveillance operatives will be observing the principal's surroundings, not them. As such, a quick hand gesture can be missed. Superior tradecraft calls putting on or taking off a scarf, or rather than carrying a satchel bag over the shoulder, swapping to carrying it by hand. These actions are longer in duration, so they are less likely to be missed by the principal's countersurveillance team.

Because the principal is not required to conduct their surveillance detection, there may be times whether they are heading toward or are in danger. During times like this, the countersurveillance team needs to signal the principal to take some other form of action. Predetermined action and corresponding signals would be part of the SMEAC briefing. An example might be to move ahead of the principal and place an object in their path—say, an unusual fast-food wrapper (which is an item that can be easily carried folded.

---

79. Countersurveillance operatives may carry radio frequency scanners to intercept any two-way communication of the opposing surveillance team. These intercepts may also be relayed to the team's forward command post for conversion into intelligence.

ABOUT THE AUTHOR

Dr Henry (Hank) Prunckun, BSc, MSocSc, MPhil, PhD, is a former Australian government intelligence analyst. He spent much of his twenty-eight-year career in operational areas, including security, investigation, and counter-terrorism. During this time, he was conferred with two literary awards and a professional service award by the International Association of Law Enforcement Intelligence Analysts. After leaving government service, he served as a freelance private investigator and spent over a decade as a research criminologist at Charles Sturt University, Sydney, specializing in the study of transnational crime—espionage, terrorism, drugs and arms trafficking, and cyber-crime.

INDEX

www.ingramcontent.com/pod-product-compliance
Lightning Source LLC
LaVergne TN
LVHW051020080826
845145LV00009B/2719

* 9 7 8 0 6 4 5 2 3 6 2 0 0 *